Soil Erosion: Quiet Crisis in the World Economy

Lester R. Brown
and
Edward C. Wolf

Worldwatch Paper 60
September 1984

ISBN 0-916468-60-7

Printed on recycled paper

Table of Contents

Introduction 5
The Causes of Soil Erosion 9
Dimensions of the Problem 14
The Erosion of Productivity 23
Erosion's Indirect Costs 27
The Economics of Conserving Soil 31
The Governmental Role 35
The Global Balance Sheet 40
Notes 45

Introduction

5

Over the past generation world food output has more than doubled. Coming at a time when little new land was brought under the plow, this was an impressive achievement. But we can now see that this remarkable feat has a high price: Some of the agricultural practices that boosted food production have also led to excessive soil erosion.

Spurred by both population growth and rising affluence, world demand for food climbs higher each year. In the face of this continuously expanding demand and the associated relentless increase in pressures on land, soil erosion is accelerating. Anson R. Bertrand, a senior official of the U.S. Department of Agriculture, has described the situation in the United States: "The economic pressure—to generate export earnings, to strengthen the balance of payments and thus the dollar—has been transmitted more or less directly to our natural resource base. As a result soil erosion today can be described as epidemic in proportion."[1]

Bertrand's perceptive, though sobering, linkage between economic forces and soil deterioration applies to other countries as well. Most countries feel pressure to feed their own people, rather than to expand food exports. But demands placed on soils are increasing worldwide. Each year the world's farmers must now attempt to feed 81 million more people, good weather or bad.[2]

Soil erosion is a natural process, one that is as old as the earth itself. But today soil erosion has increased to the point where it far exceeds the natural formation of new soil. As the demand for food climbs, the world is beginning to mine its soils, converting a renewable resource

This paper is an updated version of the chapter "Conserving Soils" that appeared in *State of the World 1984* (W. W. Norton & Co., 1984).

into a nonrenewable one. Even in an agriculturally sophisticated country like the United States, the loss of soil through erosion exceeds tolerable levels on some 44 percent of the cropland. Indeed, the U.S. crop surpluses of the early eighties, which are sometimes cited as the sign of a healthy agriculture, are partly the product of mining soils.

The incessant growth in demand for agricultural products contributes to soil erosion in many ways. Throughout the Third World farmers are pushed onto steeply sloping, erosive land that is rapidly losing its topsoil. Elsewhere, such as the American Midwest, many farmers have abandoned ecologically stable, long-term rotations, including hay and grass, as well as row crops, in favor of the continuous row cropping of corn or other crops. In other areas farming has extended into semiarid regions where land is vulnerable to wind erosion when plowed.

The loss of topsoil affects the ability to grow food in two ways. It reduces the inherent productivity of land, both through the loss of nutrients and degradation of the physical structure. It also increases the costs of food production. When farmers lose topsoil they may increase land productivity by substituting energy in the form of fertilizer, or through irrigation to offset the soil's declining water absorptive capacity. Farmers losing topsoil may experience either a loss in land productivity or a rise in costs. But if productivity drops too low or costs rise too high, farmers are forced to abandon their land.

Grave though the loss of topsoil may be, it is a quiet crisis, one that is not widely perceived. And unlike earthquakes, volcanic eruptions or other natural disasters, this humanmade disaster is unfolding gradually. Often the very practices that cause excessive erosion in the long run, such as the intensification of cropping patterns and the plowing of marginal land, lead to short-term production gains, creating an illusion of progress and a false sense of security.

Over most of the earth's surface, the thin mantle of topsoil on which agriculture depends is six to eight inches thick. Although the depletion of this thin layer of soil may compromise economic progress and

"Nowhere has the depletion of topsoil gained the attention paid to the depletion of oil reserves."

political stability even more than dwindling oil reserves, nowhere has the depletion of topsoil gained the attention paid to the depletion of oil reserves. Fifteen years ago, the public was largely unaware of the rate of oil depletion, but that changed with the oil price hikes of 1973 and 1979.

Governments everywhere have responded to the growing scarcity of oil, but such is not the case for soil. With only occasional exceptions, national agricultural and population policies have failed to take soil depletion into account. In part, the contrasting awareness of oil and soil depletion is the understandable product of differing levels of information. Estimates are regularly made for oil reserves, adjusting annually both for depletion through production and for new discoveries. Such a procedure does not exist for world soil reserves. Indeed, not until topsoil has largely disappeared and food shortages have developed or famine threatens does this loss become apparent.

Projections of the world supply and demand of food made in the late sixties and early seventies did not anticipate the slowdown in growth of world food output over the last decade. Nor did they anticipate the sustained decline since 1970 of per capita food production in Africa and the Andean countries of Latin America. One reason for this shortcoming may have been the failure to incorporate the effect of soil erosion on food production. Projections of world food production always incorporate estimates of future cropland area, but what has been lacking thus far has been any effort to project changes in inherent productivity of the projected cropland area.

To help remedy this shortcoming in world food supply projections we have undertaken an estimate of the worldwide loss of topsoil from cropland. Making such an estimate is not a simple matter of tabulating data from individual countries; unfortunately, few have attempted to measure their topsoil loss. We do know that if the current rate of topsoil loss through erosion continues there will be a sharp reduction in the amount of topsoil available by the end of the century. Largely because of population growth, we project a 19 percent decline in cropland per person between 1984 and the end of the century. But our projections of the amount of topsoil per person, assuming current rates of soil erosion continue, shows a decline of 32 percent from 1984 to 2000. The difference between the decline in cropland area per person and topsoil per person is significant, and this difference will affect food production trends.[3]

The United States is the only major food producer that has taken a systematic inventory of its topsoil. The Soil Conservation Service of the Department of Agriculture assesses soil resources and rates of erosion every five years. The first assessment was completed in 1977, the second in 1982. Data from the two studies indicates little change in the rate of excessive erosion of some 1.7 billion tons per year over this five-year span.[4]

Any estimate of world soil losses depends on fragments of data that exist for various parts of the world, as well as assumptions to fill in the gaps in the global data fabric. Our estimates draw on many sources. Direct evidence includes national soil surveys compiled for some countries; local studies by soil scientists within various countries; data from United Nations agencies such as the Food and Agriculture Organization and the United Nations Environment Program; and studies, both published and unpublished, from international aid agencies including the World Bank, the Agency for International Development and the Swedish International Development Authority. More indirect sources of evidence include hydrological studies of river silt loads; rates of reservoir sedimentation; meteorological studies of atmospheric dust flows; private communication with soil scientists and government officials throughout the world; and lastly, personal experiences and observations when traveling.

Our estimate of world topsoil loss from cropland is not highly refined and by no means final. Though inexact, it is presented here to draw attention to a process that will eventually undermine the world economy if not arrested. Without some sense of how fast soils are being lost, it will be difficult to mobilize the resources to save them.

The Causes of Soil Erosion

9

The apparent increase in soil erosion over the past generation is not the result of a decline in the skills of farmers but rather of the pressures on farmers to produce more. In an integrated world food economy, the pressures on land resources are not confined to particular countries; they permeate the entire world. Many traditional agricultural systems that were ecologically stable as recently as mid-century, when there were only 2.5 billion people in the world, are breaking down as world population moves toward 5 billion.

Over the millennia, as the demand for food pressed against available supplies farmers devised ingenious techniques for extending agriculture onto land that was otherwise unproductive while still keeping erosion in check and maintaining land productivity. These techniques include terracing, crop rotations, and fallowing. Today, land farmed through these specialized techniques still feeds much of humanity. Although these practices have withstood the test of time, they are breaking down in some situations under the pressure of continuously rising demand.

In mountainous regions such as those in Japan, China, Nepal, Indonesia, and the Andean countries, construction of terraces historically permitted farmers to cultivate steeply sloping land that would otherwise quickly lose its topsoil. Centuries of laborious effort are embodied in the elaborate systems of terraces in older settled countries. Now the growing competition for cropland in many of these regions is forcing farmers up the slopes at a pace that does not permit the disciplined construction of terraces of the sort their ancestors built, when population growth was negligible by comparison. Hastily constructed terraces on the upper slopes often begin to give way. These in turn contribute to landslides that sometimes destroy entire villages, exacting a heavy human toll. For many residents of mountainous areas in the Himalayas and the Andes, fear of these landslides has become an integral part of daily life.

Research in Nigeria has shown how much more serious erosion can be on sloping land that is unprotected by terraces. Cassava planted on land of a 1 percent slope lost an average of 3 metric tons per hectare each year, comfortably below the rate of soil loss tolerance. On a 5 percent slope, however, land planted to cassava eroded at a rate of 87 tons per hectare annually—a rate at which a topsoil layer of six inches would disappear entirely within a generation. Cassava planted on a 15 percent slope led to an annual erosion rate of 221 tons per hectare, which would remove all topsoil within a decade. Intercropping cassava and corn reduced soil losses somewhat, but the relationship of soil loss and slope remained the same.[5]

Throughout the Third World increasing population pressure and the accelerating loss of topsoil seem to go hand in hand. Soil scientists S. A. El-Swaify and E. W. Dangler have observed that it is in precisely those regions with high population density that "farming of marginal hilly lands is a hazardous necessity. Ironically, it is also in those very regions where the greatest need exists to protect the rapidly diminishing or degrading soil resources." It is this vicious cycle, set in motion by the growing human demands for food, feed, fiber, and energy, that makes mounting an effective response particularly difficult.[6]

In other parts of the world farmers have been able to cultivate rolling land without losing excessive amounts of topsoil by using crop rotations. Typical of these regions is the midwestern United States, where farmers traditionally used long-term rotations of hay, pasture, and corn. Fields planted in row crops, such as corn, are most susceptible to erosion. By alternating row crops with cover crops like hay, the average annual rate of soil erosion was kept to a tolerable level. Not only do crop rotations provide more soil cover, but the amount of organic matter that binds soil particles together remains much higher than it would under continuous row cropping.

As world demand for U.S. feedstuffs soared after World War II and as cheap nitrogen fertilizer reduced the need for legumes, American farmers throughout the Midwest, the lower Mississippi Valley, and the Southeast abandoned crop rotations to grow corn or soybeans

"Throughout the Third World increasing population pressure and the accelerating loss of topsoil seem to go hand in hand."

continuously. The risks associated with this shift in cropping patterns have long been known. Research undertaken in Missouri during the thirties showed an increase in soil erosion from 2.7 tons per acre (1 acre equals 0.4 hectares) annually when land was in a corn-wheat-clover rotation to 19.7 tons per acre when the same land was planted continuously to corn. (See Table 1.) The lower rate is within established erosion tolerance levels, whereas the higher rate would lead to the loss of one inch of topsoil in less than a decade. Much of the decline in inherent soil fertility that occurs under row crops is being masked by advances in technology, particularly by the increasing use of chemical fertilizer.

Fallowing has permitted farmers to work the land both in semiarid regions and in the tropics, where nutrients are scarce. In vast semi-arid areas—such as Australia, the western Great Plains of North America, the Anatolian plateau of Turkey, and the drylands of the Soviet Union—where there is not enough moisture to support continuous cultivation, alternate-year cropping has evolved. Under this system land is left fallow without a cover crop every other year to accumulate moisture. The crop produced in the next season draws on two years of collected moisture.

In some situations this practice would lead to serious wind erosion if strip-cropping were not practiced simultaneously. Alternate strips

Table 1: Cropping Systems and Soil Erosion

Cropping System	**Average Annual Loss of Soil**
	(tons/acre)
Corn, wheat, and clover rotation	2.7
Continuous wheat	10.1
Continuous corn	19.7

Source: M. F. Miller, "Cropping Systems in Erosion Control," Missouri Experiment Station, Bulletin 366 (1936), reprinted in National Agricultural Lands Study, *Soil Degradation: Effects on Agricultural Productivity* (Washington, D.C.: 1980).

planted to crops each year serve as windbreaks for the fallow strips. This combination of fallowing and strip-cropping permitted wheat production to continue in the western U.S. Great Plains after the Dust Bowl years.[7]

Rising demand for food has reduced the area fallowed in key dryland farming regions. As world wheat prices climbed sharply during the mid-seventies, U.S. summer fallow land dropped from 17 million hectares in 1969 to 13 million hectares in 1974.[8] This decline led Kenneth Grant, head of the U.S. Soil Conservation Service, to warn farmers that severe wind erosion and dust bowl conditions could result. He cautioned farmers against the lure of record wheat prices and short-term gains that would sacrifice the long-term productivity of their land. By 1977, the National Resources Inventory showed that wind erosion in wheat-growing states such as Texas and Colorado far exceeded the tolerance levels.[9]

At the same time, the amount of fallowed land in the Soviet Union was also being reduced. During the late sixties and early seventies, the Soviets were consistently fallowing 17-18 million hectares each year in the dryland regions. But after the massive crop shortfall and heavy imports of 1972, the fallowed area was reduced by one-third. By the early eighties Soviet officials were returning more land to fallow in an effort to restore land productivity.[10]

In the tropics—such as parts of Africa south of the Sahara, Venezuela, the Amazon Basin, and the outer islands of Indonesia—fallowing is used to restore the fertility of the soil. In these areas more nutrients are stored in vegetation than in the soil. When cultivated and stripped of their dense vegetative cover, soils of the humid tropics quickly lose their fertility. In response to these conditions, farmers have evolved a system of shifting cultivation: They clear and crop land for two, three, or possibly four years and then systematically abandon it as crop yields decline. Natural vegetation soon takes over the abandoned field. Moving on to fresh terrain, farmers repeat the process. When these cultivators return to their starting point after 20-25 years, the soil has regained enough fertility to support crop production for a few years.

Mounting population pressures in the tropics are forcing shifting cultivators to shorten these rotation cycles. As this happens, land productivity falls. A 1974 World Bank study reported that in Nigeria "fallow periods under shifting cultivation have become too short to restore fertility in some areas."[11] In some locales the original cropping cycle of 10-15 years has already been reduced to 5. Since 1950, the cropped area in Nigeria has multiplied 2.5 times as new land, largely marginal, has been added and fallow cycles have been shortened. Together, these two trends have offset the gains from the increased use of chemical fertilizer, the adoption of improved varieties, and the expansion of irrigation. Cereal yields are no higher than they were in the early sixties.

As population pressure has intensified in the river floodplains of northern Thailand over the last few decades, rice farmers have migrated to the nearby uplands where they practice rain-fed cropping. Early migrants adopted a "slash and burn" system with a cycle of eight to ten years, which seemed quite stable. As migration has continued, however, the forest fallow cycle has been shortened to between two to four years in many areas. In analyzing this situation, John M. Schiller concluded that "soil erosion problems are becoming clearly manifest in some areas and the effect of increased runoff is causing increased flooding in the lowlands and the siltation of dams. A potentially very unstable physical, economic, and social situation is developing in the affected areas."[12]

Similar pressures on land are evident in tropical Latin America. According to U.N. Food and Agriculture Organization researchers: "There is abundant evidence in certain regions of Venezuela that, with growing population pressure, the fallow period is becoming increasingly shorter so that soil fertility is not restored before recropping. This leads to a fall in the organic content and the water holding capacity of the soil. Soil structure deteriorates and compaction becomes more common . . . in other words, with the population of modern times, formerly stable, shifting cultivation systems are now in a state of breakdown."[13]

Another source of accelerated soil erosion in recent years has been the shift to larger farm equipment, particularly in the Soviet Union and United States. In the United States, for example, the shift to large-scale equipment has often led to the abandonment of field terraces constructed to reduce runoff on sloping lands. In dryland farming regions, tree shelter belts that interfere with the use of large-scale equipment have also been removed. The enlargement of fields to accommodate huge tractors and grain combines also reduces border areas that have traditionally served as checks on erosion.

This transformation of agricultural practices has been fueled by the growing worldwide demand for U.S. feed crops, particularly corn and soybeans, and by the availability of cheap chemical fertilizer. Demand growth, in turn, has been amplified by population growth that has hastened the deterioration of traditional agriculture in many countries. As a result, agricultural systems throughout the world are now experiencing unsustainable levels of soil loss.

Dimensions of the Problem

One of the first scientists to assess the dimensions of world soil erosion was geologist Sheldon Judson, who estimated in 1968 that the amount of river-borne soil sediment carried into the oceans had increased from 9 billion tons per year before the introduction of agriculture, grazing, and other activities to 24 billion tons per year. Judson observed: "There is no question that man's occupancy of the land increases the rate of erosion. Where that occupation is intense and is directed to the use of land for cultivated crops, the difference is one or more magnitudes greater than when the land is under a complete natural vegetative cover, such as grass or forest." His estimates indicate that humans have become an important geologic agent, accelerating the flow of soil to the oceans.[14]

Although detailed information on soil erosion at the local level is available for only a few countries, data on the sediment load of the world's major rivers and on the wind-borne movement of soil over the oceans do provide a broad-brush view of soil erosion at the

"The Ganges of India deposits 1.5 billion tons of soil into the Bay of Bengal every year."

Table 2: Sediment Load of Selected Major Rivers

River	**Countries**	**Annual Sediment Load**
		(million metric tons)
Yellow	China	1,600
Ganges	India	1,455
Amazon	several	363
Mississippi	United States	300
Irrawaddy	Burma	299
Kosi	India	172
Mekong	several	170
Nile	several	111

Source: S. A. El-Swaify and E. W. Dangler, "Rainfall Erosion in the Tropics: A State of the Art," in American Society of Agronomy, *Soil Erosion and Conservation in the Tropics* (Madison, Wisc.: 1982).

continental level. The most recent figures on river sediment flow show the world's major rivers carrying heavy loads of soil to the oceans. Data compiled in 1980 by three Chinese scientists working for the Yellow River Conservancy Commission in Beijing indicated that river was carrying 1.6 billion tons of soil to the ocean each year. (See Table 2.) Hydrologists estimate that on average one-fourth of the soil lost through erosion in a river's watershed actually makes it to the ocean as sediment. The other three-fourths is deposited on foot-slopes, in reservoirs, on river floodplains or other low-lying areas, or in the riverbed itself, which often causes channel shifts.[15]

Close behind the Yellow River, in terms of silt load, is the Ganges of India, which deposits 1.5 billion tons of soil into the Bay of Bengal each year. The Mississippi, the largest U.S. river, carries 300 million tons of soil into the Gulf of Mexico each year, far less than the Yellow or the Ganges. Yet it represents topsoil from the agricultural heartland and is thus a source of major concern for U.S. agronomists.[16]

Scientists have recently documented that vast amounts of wind-borne soil are also being deposited in the oceans as sediment. Island-based air sampling stations in the Atlantic, along with recent satellite photographs, indicate clearly that large quantities of soil dust are being carried out of North Africa over the Atlantic. Visible from satellites, these huge plumes of fine soil particles from the arid and desert expanses of North Africa at times create a dense haze over the eastern Atlantic. Estimates of the amount of African soil being carried west in this way, reported in four studies between 1972 and 1981, range from 100-400 million tons annually, with the latest report being at the upper end of the range.[17]

A 1983 *Science* article reported a similar loss of soil from Asia, soil that is carried eastward over the Pacific. Air samples taken at the Mauna Loa Observatory in Hawaii from 1974 through 1982 indicate a continuous movement of soil particles from the Asian mainland, with a peak annual flow consistently occurring in March, April, and May, a time that coincides with a period of strong winds, low rainfall, and plowing in the semiarid regions of North Asia. Scientists at Mauna Loa can now tell when spring plowing starts in North China.[18]

Although soil erosion data are not available for most countries, a rough estimate of the excessive worldwide loss of topsoil from croplands is needed. Without such an estimate, assessments of the world food prospect are unrealistic. The estimate developed in the following pages is the best that we can construct from the information now available. If other governments were to follow the U.S. lead and take careful inventories of their soil resources to determine the rate of excessive erosion, then they would have the information needed to formulate realistic agricultural and population policies.

The United States is one of the few countries that has analyzed soil losses in detail. As directed by the Rural Development Act of 1972, the Soil Conservation Service undertook an exhaustive inventory of land use and soil loss in 1977. Based on some 200,000 data samplings, it yielded remarkably detailed information on local soil loss throughout the United States. The first inventory formed the basis for the comprehensive appraisal of the nation's soil and water resources mandated by the Soil and Water Resources Conservation Act of 1977. A second inventory, conducted in 1982, expanded the survey to

"Scientists at Mauna Loa (Hawaii) can now tell when spring plowing starts in North China."

nearly one million sample points nationwide, the most comprehensive look at soil resources ever completed by any country.[19]

Rates of soil loss revealed by the National Resources Inventory can be related roughly to the tolerable level, a rate that would not impair long-term productivity. Calculated at from one to five tons per acre annually, depending on soil and climatic conditions, this figure represents the maximum level of soil erosion that will permit a high level of crop productivity to be sustained economically and indefinitely. The 1982 inventory showed that 44 percent of U.S. cropland was losing topsoil in excess of its soil loss tolerance level. The loss of soil at this excessive rate from the U.S. cropland base of 421 million acres totaled 1.7 billion tons, with over 90 percent coming from less than one-quarter of the cropland.[20]

India is one of the few other countries to compile a national estimate of soil loss. In 1975, Indian agricultural scientists collected data on local soil erosion from each of the research stations in the national network maintained by the Indian Council for Agricultural Research. Using these figures, they estimated that 6 billion tons of soil are eroded from India's croplands each year.[21] From this and from an estimate that 60 percent of the cropland is eroding excessively, the excessive topsoil loss can be calculated by subtracting from the total a tolerance level of five tons per acre. This yields an excessive topsoil loss from Indian cropland of 4.7 billion tons per year, more than twice the U.S. level. This estimate rests on far less data than does the figure for the United States but it is based on information from agricultural scientists familiar with local soils and it is corroborated by data on siltation of hydroelectric reservoirs, river sediment loads, and other indirect indicators.

The Soviet Union, which has the world's largest cropland area, may be losing more topsoil than any other country. Although detailed information on the extent of the loss is not available, numerous sources—including Soviet research reports, public statements by scientists and government officials, and the observations of visitors from abroad—indicate the severity of the problem. Papers published by the Soil Erosion Laboratory at the University of Moscow, for example, indicate a severe and worsening erosion situation.

During the early eighties the official Soviet press carried statements by soil scientists pleading with the agricultural bureaucracy to address the loss of topsoil. And in early 1981, Dr. Vladimir Borovsky, a prominent soil scientist and director of the Kazakh Institute of Soil Science, publicly charged the Academy of Agricultural Sciences with neglect of soil problems. In a broadcast on Moscow radio, Borovsky argued that Soviet agriculture will be retarded without effective soil management. His warnings have received some support at the highest levels of Soviet government, with Politboro member Mikhail S. Gorbachev urging planners to heed the advice of soil scientists.[22] But in the face of pressures to expand production and reduce the food import deficit—now the world's largest—soil scientists are often ignored and responsible soil management practices are cast aside.

As in the United States, erosion has been spurred by the shift to large, heavy equipment and the enlargement of fields, which eliminated many natural boundary constraints on erosion of soil by both wind and water. Each year an estimated half-million hectares of cropland are abandoned because they are so severely eroded by wind that they are no longer worth farming. One scholar of Soviet environmental policies and trends, Thane Gustafson, observes, "Fifty years of neglect have left a legacy of badly damaged soils."[23]

Although there are no official figures on soil erosion, an estimate of Soviet soil losses based on the local data that are available can be compared with the situation in the United States, where detailed erosion information has been collected. Two Soviet scientists, P. Poletayev and S. Yashukova, writing on environmental protection and agriculture in a Soviet economics journal in 1978, reported that "two-thirds of the plowed land in the Soviet Union has been subjected to the influence of various forms of erosion." Knowing the area affected by erosion, only the rate of erosion need be determined to estimate the total topsoil loss.[24]

Like the United States, the Soviet Union has an extensive dryland farming area and a substantial irrigated area. The European Soviet Union, which accounts for a large share of total farm output, has moisture levels similar to the U.S. Midwest. In terms of rainfall intensity, topography, and erodibility of prevailing soil types, nothing indicates that soil erosion in the Soviet Union would be markedly less than in the United States. Where cropping patterns are concerned, the Soviet Union relies much more heavily on small grains, whereas the United States relies relatively more on row crops, such as corn and soybeans.

Much of the Soviet grain land, however, remains bare during the winter and early spring, when rainfall is heaviest in many regions of the country. In a paper presented in the United States in 1983, P.S. Tregubov of the Dokuchaev Soil Institute in Moscow reported that land left in bare fallow to be sown to winter crops sustained losses that far exceeded the rate of new soil formation. He observed that "spring was found to be the most dangerous period because soils are characterized by fluidity after snow thawing." To document this, Tregubov cited long-term experiments showing a mean annual soil loss on bare fallow of 59 tons per hectare annually in the Baltic Sea shore regions, 46 tons per hectare in the Rostov region, and 32 tons per hectare in the Transcaucasian region. By comparison, in the American states with the most severe erosion rates in 1982, Texas lost nearly 40 tons per hectare, Colorado lost 32 tons, and Iowa, just over 30 tons.[25]

These data and observations suggest it is not unreasonable to assume that Soviet soils are eroding at least as rapidly as those in the United States. If 44 percent of the land is affected by erosion at the same rate as in the United States, which may be a conservative assumption, the excessive loss of topsoil from Soviet croplands is over 2.5 billion tons per year.

In China, the fourth major food-producing country, river siltation is now a nationally recognized threat—one that has reached dimensions unmatched elsewhere. Dust storms in the north and the siltation of major rivers indicate the heavy soil loss. Observations by outsiders who have been called in to help assess soil conditions indicate that the erosion rate in China is at least as great as that in India, where more detailed data are available.

Table 3: Observations of Soil Erosion in the Third World

Country	Observation	Source
Nepal (Katmandu)	"Local inhabitants . . . all concur that the problem is more severe now than a generation ago."	*Mountain Research and Development*, (Boulder, CO) 1982
Peru	"Erosion is estimated to affect between 50 and 60 percent of the surface of the whole country."	*Mountain Research and Development*, (Boulder, CO) 1982
Indonesia (Java)	"Soil erosion is creating an ecological emergency in Java, a result of overpopulation, which has led to deforestation and misuse of hillside areas by land-hungry farmers. Erosion is laying waste to land at an alarming rate, much faster than present reclamation programs can restore it."	U.S. Embassy, Jakarta, 1976
Ethiopia	"There is an environmental nightmare unfolding before our eyes . . . over 1 billion tons of topsoil flow from Ethiopia's highlands each year."	U.S. AID Mission, Addis Ababa, 1978
South Africa	"The province of Natal, incorporating Kwazulu, is losing 200 million tons of topsoil annually."	John Hanks, Institute of Natural Resources, Natal, 1980
Bolivia	"Recent aerial photographs have shown the rapid extension of desert-like conditions caused by wind erosion."	Hélène Rivière d'Arc, Institut des Hautes Études d'Amérique Latine, Paris, 1980
Iran	"The area of abandoned cultivated land has doubled in recent years."	Harold Dregne, Texas Tech University, 1971

A comparison of the sediment load of the Yellow River in China with the Ganges in India indicates the relative magnitude of soil loss through erosion faced by these two population giants. The Ganges, with a drainage basin of 1.1 million square kilometers, carries an annual sediment load of 1.46 billion tons of soil, while the Yellow River, which has a drainage basin of 668,000 square kilometers, carries 1.6 billion tons of soil to the ocean each year. These numbers suggest that the rate of soil loss in China is substantially greater than in India. For the purposes of constructing a rough global estimate, it can be assumed that the erosion rate on China's cropland exceeds India's by 30 percent. Given China's smaller cropland area, this means that China's excessive loss of topsoil from its croplands totals 4.3 billion tons per year.

For most Third World countries information on soil erosion is largely indirect, such as data on sedimentation of reservoirs and river silt loads. Other indirect sources include information on cropland abandonment as a result of severe erosion and crop reports showing long-term declines in yields. Among the most graphic sources are reports by agricultural scientists, development technicians, and other observers. (See Table 3.)

Altogether, the excessive loss of topsoil from cropland in the four major food-producing countries, which have 52 percent of the world's cropland and account for over half of its food production, is estimated at 13.2 billion tons per year. To obtain a rough idea of excessive soil erosion for the world as a whole, an assumption must be made about other countries. If the rates of soil erosion for the rest of the world are similar to those of the "big four"—which is a conservative assumption given the pressures on land in the Third World—then the world is now losing an estimated 25.4 billion tons of soil from croplands in excess of new soil formation. (See Table 4.)

Because of the shortsighted way one-third to one-half of the world's croplands are being managed, the soils on these lands have been converted from a renewable to a nonrenewable resource. Assuming

Table 4: Estimated Excessive Erosion of Topsoil From World Cropland

Country	Total Cropland	Excessive Soil Loss
	(million acres)	(million tons)
United States	421	1,700
Soviet Union	620	2,500
India	346	4,700
China	245	4,300
Total	1,632	13,200
Rest of World	1,506	12,200
Total	3,138	25,400

Source: Worldwatch Institute estimates.

an average depth of remaining topsoil of seven inches, or 1,120 tons per acre, and a total of 3.1 billion acres of cropland, there are 3.5 trillion tons of topsoil with which to produce food, feed, and fiber. At the current rate of excessive erosion, this resource is being depleted at 0.7 percent per year—7 percent each decade. In effect, the world is mining much of its cropland, treating it as a depletable resource, not unlike oil.

When most of the topsoil is lost on land where the underlying formation consists of rock or where the productivity of the subsoil is too low to make cultivation economical, it is abandoned. More commonly, however, land continues to be plowed even though most of the topsoil has been lost and even though the plow layer contains a mixture of topsoil and subsoil, with the latter dominating. Other things being equal, the real cost of food production on such land is far higher than on land where the topsoil layer remains intact.

The Erosion of Productivity

Whenever erosion begins to exceed new soil formation, the layer of topsoil becomes thinner, eventually disappearing entirely. As the topsoil layer is lost, subsoil becomes part of the tillage layer, reducing the soil's organic matter, tilth, and aeration, and adversely affecting other structural characteristics that make it ideal for plant growth. This overall deterioration in soil structure is usually accompanied by a reduced nutrient retention capacity, which lowers productivity further. Additional chemical fertilizer can often compensate for the loss of nutrients, but the deterioration of soil structure is difficult to remedy.

The effects of erosion on productivity are not easily measured since they are usually gradual and cumulative. In an effort to understand the erosion/productivity relationship better, the U.S. Secretary of Agriculture in 1980 appointed a National Soil Erosion-Soil Productivity Research Planning Committee. Among other things, committee members began gathering data from past experiments to establish an empirical foundation for predicting the effect of continuing soil loss on crop yields and production costs. They reported that when corn was grown continuously on a plot in Iowa from which the topsoil had been removed, yields were only 20 percent of those on a control plot. In an experiment in Missouri, corn yields on a desurfaced plot were 47 percent of those on the control plot. In this case, the subsoil was a clay loam—a higher-quality subsoil than is commonly the case.[26]

In an experiment in East Texas, cotton yields on land with the topsoil removed averaged only 32 percent of the control plot's. And in Minnesota, yields on severely eroded soils were roughly two-thirds those on slightly eroded soils.[27] A 1979 experiment on piedmont soils in Georgia designed to measure the effects of erosion on corn yields showed that severely eroded, moderately eroded, and uneroded soils averaged 36, 75, and 92 bushels respectively. On these soils, researchers estimated that each centimeter of topsoil lost through water erosion reduced the average corn yield by 2.34 bushels per acre.[28]

Leon Lyles, an agricultural engineer with the U.S. Department of Agriculture, has provided probably the most comprehensive collection of research results on the effect of soil erosion on land productivity. Drawing on the work of U.S. soil scientists both within and outside government, Lyles compared 14 independent studies, mostly undertaken in the corn belt states, to summarize the effects of a loss of one inch of topsoil on corn yields. His survey found that such a loss reduced yields by as little as 3 bushels per acre to as much as 6.1 bushels per acre. (See Table 5.) In percentage terms, the loss of an inch of topsoil reduced corn yields at these 14 sites by an average of 6 percent. Results for wheat, drawing on 12 studies, showed a similar relationship between soil erosion and land productivity. The loss of an inch of topsoil reduced wheat yields 0.5-2.5 bushels per acre. In percentage terms, the loss of an inch reduced wheat yields an average of 6 percent, exactly the same as for corn. (See Table 6.)

Table 5: Effect of Topsoil Loss on Corn Yields

Location	Yield Reduction Per Inch of Topsoil Lost		Soil Description
	(bushels/acre)	(percent)	
East Central, Illinois	3.7	6.5	Swygert silt loam
Fowler, Indiana	4.0	4.3	Fowler, Brookston, and Parr silt loams
Clarinda, Iowa	4.0	5.1	Marshall silt loam
Greenfield, Iowa	3.1	6.3	Shelby silt loam
Shenandoah, Iowa	6.1	5.1	Marshall silt loam
Bethany, Missouri	4.0	6.0	Shelby and Grundy silt loams
Columbus, Ohio	3.0	6.0	Celina silt loam
Wooster, Ohio	4.8	8.0	Canfield silt loam

Sources: Various reports cited in Leon Lyles, "Possible Effects of Wind Erosion on Soil Productivity," *Journal of Soil and Water Conservation,* November/December 1975.

Table 6: Effect of Topsoil Loss on Wheat Yields

Location	Yield Reduction Per Inch of Topsoil Lost		Soil Description
	(bushels/acre)	(percent)	
Akron, Colorado	0.5	2.0	Weld silt loam
Geary County, Kansas	1.3	6.2	(not available)
Manhattan, Kansas	1.1	4.3	Smolan silty clay loam
Columbus, Ohio	1.3	5.3	Cropped soil
Oregon	1.0	2.2	Deep soil
Oregon	2.5	5.8	Thin soil
Palouse area, Washington	1.6	6.9	(not available)

Sources: Various reports cited in Leon Lyles, "Possible Effects of Wind Erosion on Soil Productivity," *Journal of Soil and Water Conservation,* November/December 1975.

All the studies on soil erosion and land productivity that Lyles cited showed that the excessive loss of topsoil lowered yields measurably, although the extent of yield reduction varied. And, as noted, his compilation of studies showed a remarkable similarity in the effect of soil erosion on the yield of wheat, a crop usually grown under lower rainfall conditions, and that of corn, usually grown in areas of higher rainfall. Recent, more detailed research on three soil types in Minnesota shows that the effect of erosion varies with soil type and depth. It specifically notes that on some deeper soils, such as the Kenyon soils that are 76 inches deep, the near-term effects of erosion are negligible.[29] For the world as a whole soils of this depth constitute a small share of the total, an exception to the more typical 6-8 inches of topsoil found on most cropland.

Perhaps the most detailed analysis to date of the long-term effects of soil erosion on land productivity and food production costs is one undertaken for the Southern Iowa Conservancy District. Conducted by an interdisciplinary team of three scientists, this analysis assumed that soil erosion would continue at recent rates. The researchers classified the degree of erosion into three phases: soils that are slightly eroded, with no appreciable mixing of subsoil and topsoil in the plow layer; those that are moderately eroded, with some mixing of subsoil into the plow layer; and severely eroded soils, where the topsoil is largely gone and the plow layer is predominantly subsoil.

In 1974, the base year, 2.1 million of the district's 3.5 million acres of cropland fell into one of the three erosion phases, with the largest acreage being in the moderately eroded category. Assuming a continuation of the same rate of erosion, this would also be true for the year 2000. But by 2020, the researchers predicted, the largest share would be in the severely eroded category. As soils progress from the moderately to severely eroded category, the amount of nitrogen, phosphorus, and potash needed to grow corn increases by 38 pounds per acre. (See Table 7.) Closely paralleling this would be an increase in fuel requirements for tillage. As erosion proceeds, soils become more compact and difficult to till. The actual fuel increase varied widely by soil type, but on the average the severely eroded soils would require 38 percent more fuel for tillage than the slightly eroded soils.

Soil erosion would not only raise the costs of production by increasing the amount of fertilizer and fuel used, it would also reduce yields. For corn, a shift from slight to moderate erosion would reduce the

Table 7: Increase in Fertilizer Needs for Corn as Soil Erodes, Southern Iowa

Change in Erosion Phase	Nitrogen	Phosphate	Potash
		(pounds per acre)	
Slight to Moderate	10	2	6
Moderate to Severe	30	1	7

Source: Paul Rosenberry, Russell Knutson, and Lacy Harmon, "Predicting Effects of Soil Depletion From Erosion," *Journal of Soil and Water Conservation*, May/June 1980.

Table 8: Reduction in Yields of Key Crops as Soil Erodes, Southern Iowa

Change in Erosion Phase	Reduction in Yield Per Acre Corn	Soybeans	Oats
		(bushels)	
Slight to Moderate	16	5	9
Moderate to Severe	7	3	4

Source: Paul Rosenberry, Russell Knutson, and Lacy Harmon, "Predicting Effects of Soil Depletion From Erosion," *Journal of Soil and Water Conservation*, May/June 1980.

average corn yield by 16 bushels per acre, while going from the moderate to severe category would lower yields another 7 bushels per acre. (See Table 8.) Although the soybean yield decline was much smaller, it was proportionate, since soybean yields are roughly one-third those of corn.

Although there are few reliable data on the effect of soil erosion on land productivity for most countries, some insights into the relationship can be derived from these U.S. studies. Given the consistency of the decline in productivity across a wide range of soil types and crops, it would not be unreasonable to assume that a similar relationship between soil erosion and land productivity exists in other countries, for the basic agronomic relationships are the same. Indeed, research on West African soils shows land productivity there to be even more sensitive to topsoil loss than in North America. The loss of 3.9 inches of topsoil in West Africa cut corn yields by 52 percent. Yields of cow peas, a leguminous crop, were reduced by 38 percent. This marked decline may attest to the fragility of tropical soils.[30]

Erosion's Indirect Costs

When farmers lose topsoil they pay for it in reduced soil fertility, but unfortunately the costs of erosion are not confined to the farm alone.

As soil is carried from the farm by runoff, it may end up in local streams, rivers, canals, or irrigation and hydroelectric reservoirs. The loss of topsoil that reduces land productivity may also reduce irrigation, electrical generation, and the navigability of waterways.

The increase in the amount of irrigated land in the world went hand in hand with efforts to raise food supplies during the third quarter of this century. Often the centerpiece of national development strategies throughout the Third World, multipurpose dams represented enormous investments and an important part of the capital stock of new nations. Typical of these was the Mangla Reservoir in Pakistan. The designers of the reservoir projected a life expectancy for the dam of at least a century. What they did not reckon on was the effect of mounting population pressure on the watershed feeding the reservoir. A combination of the axe and the plow, as land-hungry peasants push up the hillsides, is leading to a rate of siltation that will probably fill the reservoir with silt at least 25 years earlier than projected. (See Table 9.) One recent estimate predicts it will be filled within half a century.[31]

Table 9: Siltation Rates in Selected Reservoirs

Country	Reservoir	Annual Siltation Rate	Time To Fill With Silt
		(metric tons)	(years)
Egypt	Aswan High Dam	139,000,000	100
Pakistan	Mangla	3,700,000	75
Philippines	Ambuklao	5,800	32
Tanzania	Matumbulu	19,800	30
Tanzania	Kisongo	3,400	15

Source: S. A. El-Swaify and E. W. Dangler, "Rainfall Erosion in the Tropics: A State of the Art," in American Society of Agronomy, *Soil Erosion and Conservation in the Tropics* (Madison, Wisc.: 1982).

"The loss of topsoil may also reduce irrigation, electrical generation, and the navigability of waterways."

In the Philippines, scores of hydroelectric and irrigation reservoirs have been constructed, many of them with assistance from international development agencies. Here, as in Pakistan, the combination of watershed deforestation and steep slopes being cleared for cultivation is yielding record siltation rates. A report of the Agency for International Development on the prospects for the Ambuklao Dam notes that "the cutting of timber and the subsequent loss of water retention capacity of land surrounding the reservoir has resulted in massive silting of the reservoir, reducing its useful life from 60 to 32 years."[32]

One reason for the excessively rapid siltation rates is that multipurpose dams are designed by engineers who sometimes fail to recognize the impoundments they build as part of a watershed, which often drains an area of several thousand square miles. The Anchicaya Dam in Colombia is a classic example. Engineers expressed little concern with the siltation problem, even though when the project began farmers were already invading the upper reaches of the watershed that feeds the dam. Within two years of its completion, the dam had already lost a quarter of its storage capacity because of siltation.[33]

In India, the indirect costs of water-eroded soil are summed up well by B. B. Vohra, Chairman of the National Committee on Environmental Planning. He observes that the "premature siltation of our 500,000 odd ponds and of the 487 reservoirs of our major and medium irrigation and multipurpuse projects on which the community has invested over 100 billion rupees during the last three decades is a particularly serious matter." He notes that siltation rates are now commonly several times as high as the rate that was assumed when the projects were designed. (See Table 10.) Vohra observes that not only is the life expectancy of these projects being severely reduced, but "in most cases there will be no alternative sites for dams once the existing ones are rendered useless." A dam site is often unique. Once lost, it cannot be replaced. For India, what is at stake, according to Vohra, "is the loss of the irreplaceable potential—for irrigation, for electricity and for flood control—that these storages represent."[34]

Table 10: India: Siltation Rates in Selected Reservoirs

Reservoir	Assumed Rate	Observed Rate	Ratio of Observed to Assumed Erosion
	(in acre-feet)		
Bhakar	23,000	33,475	1.46
Maithon	684	5,980	8.74
Mavurakshi	538	2,000	3.72
Nizam Sugar	530	8,725	16.46
Panchet	1,982	9,533	4.81
Ramganga	1,089	4,366	4.01
Tungabhadra	9,796	41,058	4.19
Ukai	7,448	21,758	2.92

Sources: Adapted from S. A. El-Swaify and E. W. Dangler, "Rainfall Erosion in the Tropics: A State of the Art," in American Society of Agronomy, *Soil Erosion and Conservation in the Tropics* (Madison, Wisc.: 1982), and Center for Science and Environment, *The State of India's Environment 1982* (New Delhi: 1982).

The list of countries with soil-silting disasters goes on and on. The names change but the conditions are common. Whether in Nigeria, Indonesia, Pakistan, or Mexico, the same basic principles of soil physics are at work. When soil on sloping land is farmed improperly, it begins to move under the impact of rain and ends up in places where it usually does more harm than good.

The third major indirect cost of soil erosion is the loss of navigability. Perhaps the most dramatic case occurs in the Panama Canal. The combination of deforestation and the plowing of steeply sloping land in the watershed area by landless *campesinos* is leading to an unprecedented siltation of the lakes that make up part of the Canal. If the trends of the late seventies and early eighties continue, the capacity of the Panama Canal to handle shipping will be greatly reduced by the end of the century, forcing many ocean-going freighters that have relied on its 10,000 mile shortcut to make the trip via Cape Horn.[35]

Within the United States, soil once used in the Midwest to grow corn now clogs the Mississippi waterways. One of the largest items in the budget of the Army Corps of Engineers is the dredging of inland waterways, particularly in the lower Mississippi River. Vast quantities of soil reach the Gulf of Mexico to become ocean sediment, but substantial amounts are deposited on the way, making large-scale dredging imperative if this major artery connecting U.S. farms with world markets is to continue to function.

The Economics of Conserving Soil

Recent U.S. studies have rather consistently concluded that soil erosion control is often not economical for farmers, based on strictly dollars-and-cents criteria. The study of southern Iowa soils referred to earlier showed that the short-term cost to farmers of reducing soil erosion to a level that would not reduce inherent productivity would be three times as great as the benefits.

Narrow profit margins, such as those confronting U.S. farmers during the early eighties, might well mean that if farmers were to invest in appropriate conservation measures their profit margins would disappear entirely, forcing them to operate at a deficit. They would then face the prospect of bankruptcy in the near future. Alternatively, they could continue to follow existing agricultural practices and avoid near-term bankruptcy, but face the prospect of declining productivity over the long term and the eventual abandonment of land, if not by this generation then by the next. In the absence of a governmental cost-sharing program similar to those used so effectively in the past, a farmer's only choice is whether to go out of business sooner or later.

The economics of erosion control in the United States has recently become more attractive with the adoption of new minimum-tillage practices. In traditional tillage the moldboard plow, the principal farm implement, was used to turn over all the soil in seedbed preparation. With minimum tillage the land is not plowed in this traditional way. Crop residues are left on the surface and seeds are drilled directly into the unplowed land, with herbicides providing the weed control that mechanical cultivators previously did. The availability of herbicides enabled minimum tillage in the same way that the introduction of cheap chemical fertilizer permitted farmers to dispense with crop rotation containing legumes. More commonly, however, in minimum tillage narrow strips of land are tilled where the corn, soybeans, or other row crops are planted, leaving the space between the rows undisturbed.[36]

To conserve fuel, cost-conscious U.S. farmers were already experimenting with various reduced tillage practices in the early seventies. The oil price jump in 1973 reinforced this change. The fortuitous nature of this development lies in its effect on soil erosion. With land not being plowed and crop residues left on the surface, soil is protected from rain, and runoff is much less. The result is that farmers adopting minimum-till or no-till practices to conserve fuel have discovered that they are also conserving their soil. Individual farmers doing their own cost-benefit calculations could weigh the importance of both energy and soil savings in deciding whether to adopt the new practices.

Not all soils and not all situations lend themselves to minimum tillage, which increases soil moisture and slows soil temperature rise in the spring. In areas such as the northern corn belt, the short growing season could restrict the adoption of minimum tillage. Still, the growth in reduced tillage acreage in the United States has been remarkably steady, increasing every year since data collection began in 1972. In that year, nearly 30 million acres—roughly one-tenth of the cultivated area—was in reduced tillage. (See Table 11.) In 1984, it reached 108 million acres, nearly one-third of all the land in crops.

Despite the encouraging increase in the use of minimum tillage, there is little information to date on its impact on severely eroding land. While noting the encouraging spread of reduced tillage farming practices, Robert Gray of the American Farmland Trust observes that even minimum-till and no-till practices should be forsworn on the most severely eroding land, which should not be in production at all.[37] Preliminary observations indicate that minimum tillage is being adopted by the more progressive, innovative farmers and that all too

"Farmers adopting no-till practices to conserve fuel are also conserving their soil."

Table 11: United States: Conservation Tillage, 1972-84

Year	Area in Conservation Tillage	Share of Harvested Cropland[1]
	(million acres)	(percent)
1972	29.7	10.0
1973	43.9	13.7
1974	46.7	14.2
1975	56.2	16.7
1976	59.6	17.6
1977	70.0	20.2
1978	74.8	22.2
1979	85.2	24.6
1980	88.5	25.1
1981	99.0	27.5
1982	111.9	31.2
1983	91.3[2]	30.4[2]
1984	107.5[3]	32.6[3]

[1]Share of harvested cropland, calculated by Worldwatch, includes area harvested and that on which crops failed. [2]The decline in acreage in 1983 reflects the area of cropland set aside in the Payment-in-Kind program. [3]Estimates.

Sources: *No-Till Farmer*, March 1982, March 1983, and March 1984; Worldwatch Institute.

often these are not the ones on marginal lands, which are eroding most severely. Nonetheless, the overall effect of minimum tillage on soil conservation must be viewed as positive and hopeful.

Although roughly one-third of U.S. cropland was minimum-tilled as of 1984, the practice is not widely used in other parts of the world, though there are occasional references to it. For example, the two

Soviet scientists cited earlier, who reported on environmental protection and agriculture, noted that "a method in which the soil layer is not turned in the treatment of the soil has been developed and has been widely introduced in production in Kazakhstan, the Altei, in the Urals, in the Lower Volga and other regions." They also indicated that the specialized farm equipment needed for minimum tillage is being manufactured in the Soviet Union.[38]

Some observers claim that erosion control is particularly uneconomical in the United States because of the untrammeled pursuit of profits. But all indications are that production quotas, as used in the Soviet Union for example, are at least as destructive of soils as the profit motive. Under the Soviet system, farm managers are judged not by how effectively they control soil erosion but by how successfully they fill quotas. In Western countries, landownership, particularly within a family, brings with it a certain sense of stewardship—although this obviously can be overridden by economic forces. Such a direct sense of stewardship does not exist in the Soviet Union.

Others suggest that the economics of erosion control will become clear when cropland values decline as soil erosion reduces long-term land productivity. Unfortunately, the loss of topsoil does not alter the physical appearance of the land in the short run, nor does it show up as an immediately measurable loss of productivity. Given the other factors influencing land values, such as inflation and the speculative nature of investment, the effects of soil erosion on land values are simply not visible in the short run.

Interest rates also affect the economics of erosion control. The higher interest rates are, the less attractive become investments in the soil conservation techniques, such as terracing, that pay off over the long term. Indeed, interest rates have been so high in the United States in the early eighties that farmers have simply been unable to consider seriously most investments that have a long-term payback.

> **"Production quotas, as used in the Soviet Union, are at least as destructive of soils as the profit motive."**

The Governmental Role

Although the changes in agricultural practices needed to check excessive soil erosion can usually be implemented only by farmers themselves, there are several reasons why bringing erosion under control requires government involvement. To begin with, many farmers cannot easily determine whether their erosion is excessive. Measuring the gradual loss of soil requires scientific techniques and equipment; determining whether it is excessive requires information on tolerance levels for the particular cropland in question.

Another reason for government involvement is that individual farmers may be unable to afford the conservation practices that are needed. It may make sense for society to invest in soil conservation even if it is not profitable for the individual farmer. Only government can calculate the long-term aggregate cost of soil erosion for the nation as a whole, including off-farm costs such as the silt-reduced capacity of irrigation reservoirs, hydroelectric reservoirs, or water transport systems.

The first step for governments in countries where soil erosion is believed to be a serious threat is to assess carefully the extent of soil loss. Only when such an inventory has been done can all the needed national cost-benefit calculations be made and the appropriate conservation programs designed. In India, for example, which has a cropland base comparable in size to the United States and which has only broad estimates of soil loss, it is estimated that a comprehensive nationwide soil inventory would cost some $30 million—a small price indeed compared with the contribution it could make to more intelligent policymaking.

Most governments also need better information on the relationship between soil erosion and land productivity than they now have. The National Soil Erosion-Soil Productivity Research Planning Committee in the United States notes that "such experiments are costly and time consuming. Years of data are needed to evaluate the effects of the generally slow process."[39]

The National Research Committee of the U.S. Soil Conservation Service calls work on the erosion/productivity relationship its top priority. A theoretical model called the Erosion-Productivity Impact Calculator, developed by the Department of Agriculture, is designed to evaluate both the physical process and the economic consequences of erosion. A Productivity Index that calculates the ratio between actual and potential crop yields at various levels of soil loss has been applied to soils in the major crop-producing regions of the United States and is being tested on tropical soils in Nigeria, Mexico, India, and Hawaii. Scientists coordinating the international work on the productivity index conclude that "A knowledge of the global distribution of soils combined with estimates of erosion could, using the PI, improve estimates of the global impact of erosion." But the needed inventory of the world's soils will depend on the painstaking collection of data over many years, an effort that has just begun.[40]

Mobilizing public support for adequately funded soil conservation programs will require extensive public education on the dimensions of the problem and its many consequences. Scientific proof of the necessity of soil conservation is not sufficient. Although soil scientists can chart a detailed national plan of action, as they have in the United States, they cannot call forth the political support needed to fund and administer such a plan. At this point, national political leaders must become involved.

Perhaps one of the best examples of such involvement occurred in the United States during the thirties when Franklin D. Roosevelt was President and Henry Wallace was Secretary of Agriculture. Despite the fact that the country was in the midst of perhaps the worst economic crisis in its history, Roosevelt and Wallace convinced the U.S. Congress to fund within the Department of Agriculture a new agency—the Soil Conservation Service—that would have the responsibility for administering a comprehensive soil conservation program. Roosevelt proudly took credit for the planting of tree shelter belts in the Great Plains.[41]

In a world where soil erosion is accelerating, successful national responses to this threat to food security are rare. One Third World country that has formulated an effective response is Kenya, whose program traces back to the preparation of its national statement for the 1972 U.N. Conference on the Environment held in Stockholm. When doing this national environmental assessment the Kenyan government discovered that soil erosion was by far its most serious environmental problem. Following the conference the Kenyan government requested assistance from Sweden.

In response, the Swedish International Development Authority recruited Carl-Gösta Wenner, a professor of geography specializing in soil management at the University of Stockholm, to advise the Kenyan Ministry of Agriculture. With the help of farmers in selected communities, Professor Wenner designed several local projects that were launched in 1974. Within a few years these trial efforts to conserve soil had evolved into a national program. By mid-1983 some 1,300 agricultural officers and 3,500 technical assistants had been trained in soil and water management, 50 tree nurseries had been established, and 127,000 fruit trees and 3.5 million fuel/fodder trees had been distributed to farmers. Terraces had been constructed on 100,000 farms. Farmers themselves had constructed roughly 10,000 kilometers of cutoff drains designed to reduce the erosive runoff of water. Kenyan officials proudly point out that this represents a distance equal to that from Sweden to South Africa, no small feat for a country dependent almost entirely on individual farmer initiative and labor.[42]

Professor Wenner has outlined several things that contributed to the success of the Kenyan program. First, he involved farmers in the design of the program, working with them to find out what they wanted, what they were willing to do themselves, and how much they could do with local resources. The terraces were exceedingly simple, often no more than unplowed strips of land a few feet wide across the slope. The motivation to adopt soil and water conservation practices came from the realization that yields of crops could be increased in the short-run. The demonstrable increase in crop yields associated with the terracing apparently reflected not only the saving of the soil itself, but also the retention of water and nutrients.

Aside from recognizing the need for soil conservation and organizing the program, the principal government inputs were two. The first was assistance in laying out the terraces so as to get the proper control and retention of water. And second, farmers were provided seedlings of useful trees, i.e. those that would supply fuelwood, fodder for cattle, and poles that could be used for village construction. Farmers were also given free cuttings of fodder grasses that could be established on the terraces. One advantage of having both grass that could be cut for fodder and trees that would also yield fodder was that farmers could practice stall feeding. This reduced the damage to local vegetation normally associated with freely grazing cattle and also concentrated the manure that farmers could then use to fertilize their crops.

Professor Wenner further observes that since "many farmers are more interested in fruit trees, they can be supplied with, for example, five fruit trees and 20 other trees per annum for planting around houses, along terrace edges, ditches, roads and water courses, on steep slopes and rocky ground, in shallow soil and as windbreaks."[43] The Kenyan experience demonstrates that a Third World country with limited fiscal resources and a scarcity of local skills can design and implement an effective national soil and water conservation program with a minimum of outside assistance. The key appears to be local participation in the design of a program that has demonstrable economic benefits, both short-term and long-term. The significance of this achievement goes far beyond Kenya, for it means that other countries with similar restrictions on resources can also mount an effective response, given the needed leadership.

Kenya's program also reveals the nature of the long-term commitment other governments must begin to make. By 1980, the program extension staff was able to bring conservation improvements to 30-35 thousand farms per year, a level that could stabilize Kenya's soils in 25 years. Even with staff increases, comprehensive soil conservation in Kenya would take no less than 15 years, a period during which Kenya's population is expected to nearly double.[44]

"In some countries, the expenditures required to simply stabilize soils dwarf the total appropriations for agriculture."

The ingredient missing from unsuccessful responses to the growing menace of soil erosion is political will grounded in awareness. Over the past generation, scores of countries have become food-deficit but few have linked the shortages with the depletion of their soil by erosion. In many countries people know that food prices are rising, but most don't know quite why. An understanding that lost soil means lower inherent productivity, which in turn means costlier food, is needed to inculcate a national soil conservation ethic.

In many predominantly rural societies where most people are illiterate and live at the subsistence level, a lack of public interest in soil conservation has other roots. Farmers in many Third World villages can muster little concern about the future when their immediate survival is in question. In India, reports B. B. Vohra, "an informed public opinion cannot . . . be wished into existence over night. A great deal of painstaking and patient work will have to be done to wipe out the backlog of ignorance, inertia and complacency."[45]

In much of the world today, only the willingness of national governments to share the costs of the needed measures—terracing, contour farming, strip-cropping, cover cropping, rotating crops, fallowing, and planting shelter belts—will induce farmers to fight soil erosion. One World Bank official observes that if all the Bank's capital resources of $9 billion per year were devoted to soil conservation, it would cover only a small fraction of the land affected by erosion.[46] In some countries, the expenditures required simply to stabilize soils dwarf the total appropriations for agriculture.

Within the socialist economies, where land is publicly owned and governments are directly responsible for the quality of land management, there is a need for the ruling elites to be educated about basic agronomy. Unless national political leaders understand that a country's long-term security depends on protecting its cropland, it will be difficult to get the necessary commitment of leadership time and the budgetary resources to support an effective conservation program.

The Global Balance Sheet

40 As world food demand has begun its second doubling since mid-century, pressures on land have become so intense that close to half the world's cropland is losing topsoil at a rate undermining its long-term productivity. Since agriculture is the foundation of the global economy, this loss of topsoil, if unarrested, will undermine the economy itself. Nonetheless, few countries, industrial or developing, are responding effectively to this emerging threat to economic sustainability.

Newspaper headlines that describe widening food deficits and chronic hunger in many Third World countries also describe a world finding it difficult to live within its means. Eager to maximize food output today, we are borrowing from tomorrow. The loss of over 25 billion tons of topsoil from our cropland each year is the price we pay for shortsighted agricultural policies designed to boost food output at the expense of soils, and of failed or nonexistent population policies.

In addition to the unprecedented growth in world food demand since mid-century, new demands on the world's land resources have recently emerged. When growth in the world fish catch slowed abruptly after 1970, it forced the world's consumers to turn more to land-based protein sources. With the depletion of oil reserves and the associated rise in oil prices during the seventies, the world is now turning to agriculture to produce more of the world's energy, as well as food.

Several countries have turned to agriculture as a source of liquid fuels, most importantly Brazil, which in 1984 devoted some 3.2 million acres of land to the production of sugarcane for distillation into alcohol fuel. The United States, meanwhile, used some 1.5 million acres of corn for fuel alcohol. Although this combined area is only a fraction of 1 percent of the world's cropland, it is growing steadily and should continue to do so as the transition from oil to renewable energy sources continues. Agronomist R. Neil Sampson observes that, "Seldom has such a totally new set of competitive forces been unleashed on the land as those that appear on the horizon in the declining decades of the petroleum era."[47]

The long-term social threat posed by uncontrolled soil erosion raises profound questions of intergenerational equity. If our generation persists in mining the soils so that we may eat, many of our children and their children may go hungry as a result. Agricultural economist Lloyd K. Fischer of the University of Nebraska observes that the quality of our diet in the future will be "substantially lower and the costs dramatically higher if the management of our land and water resources is not improved." He notes further that "we must cease to behave as if there were no tomorrow, or tomorrow will be bleak indeed for those who must spend their lives there."[48]

Soil erosion is a physical process, but its consequences are economic. As soils are depleted through erosion, the productivity of laborers working the eroding land becomes more difficult to raise. In agrarian societies, deterioration of this resource base makes it more difficult to raise income per person. Further, as growth in food output slows, so does overall economic output. In largely rural, low-income societies with rapid population growth, this can translate into declining per capita income, as it already has for a dozen countries in Africa.

Over the long term, world agricultural trade patterns and the international debt structure will be altered. As soils are depleted, countries are forced to import food to satisfy even minimal food needs. Scores of countries in the Third World and Eastern Europe find their international indebtedness further aggravated by their chronic dependence on imported food. And the loss of topsoil will force an energy-for-topsoil substitution as it increases the need for fertilizer and fuel for tillage. Other things being equal, land with less topsoil requires more energy to produce our food.

Soil erosion will eventually lead to higher food prices, hunger, and quite possibly, persistent pockets of famine. Although the world economy has weathered a severalfold increase in the price of oil over the past decade, it is not well equipped to cope with even modest rises in the price of food. Although the immediate effects of soil erosion are economic, the ultimate effects are social. When soils are depleted and crops are poorly nourished, people are often undernourished as well.

In efforts to conserve soil, the world is faring poorly. There are few national successes, few models that other countries can emulate. The United States has the technology, the detailed information on its soil losses and the resources to launch an exemplary soil conservation program, but it lacks leadership. Within the Third World Kenya is the only country to launch a successful national program to conserve its soils. In this respect, soil conservation contrasts sharply with oil conservation, where scores of countries have compiled impressive records in recent years. Almost everywhere dependence on petroleum is declining as it is used more efficiently. But there is no parallel with soil conservation, even though soil is a far more essential resource.

The effect of price on the depletion of soil and oil resources also varies. Higher prices for oil raise the amount that can be ultimately recovered, but higher prices for food may simply lead to more intensive land use and faster topsoil loss. And the depletion of oil reserves will make the substitution of energy for cropland more difficult, rendering the remaining soil even more valuable.

The United States unilaterally attempts to balance the world's supply and demand of agricultural commodities by withholding land from production during times of surplus. But little or no effort has been made to coordinate the farm supply management programs that divert land from production and the conservation programs designed to reduce soil erosion. U.S. farm programs have demonstrated that land can be withheld from production for supply management reasons. Unfortunately, no effort has been made to ensure that the most erosive land was set aside. As policy analyst Kenneth Cook observes, the United States has "no policy to use the good land in preference to the worst. Indeed, with respect to matching export demand to the needs of U.S. farmers and to the needs of people and resources in the developing world, we do not have responsible policy at all. We have a simple-minded sales quota."[49]

"Within the Third World
Kenya is the only country to launch
a successful national program to conserve its soils."

The United States now has an opportunity to integrate soil conservation and supply management programs. With farm program costs out of control and public support for traditional farm price support programs diminishing, Congress may be unable to legislate a new farm program in 1985 unless it directly incorporates soil and water conservation with supply management and price supports. In effect, the broad base of public support for soil conservation could be used to divert highly erosive cropland to other uses, such as fuelwood production or grazing. This would bring the production of key farm commodities down to a level that would support prices needed to make agriculture profitable. This unique opportunity for the United States ought not to be wasted. Merging the two policies, however, requires a degree of agricultural leadership that does not now exist in the United States. A recent study by the American Farmland Trust emphasizes that progress toward conserving soil awaits leaders who accept "the nondegradation of agricultural resources as a central goal of national policy." The AFT report recommends that a national strategy include a cropland reserve program for highly erodible land, an effort to cut programs that subsidize cultivation of especially fragile lands, and a reorientation of USDA technical assistance to promote cost-effective conservation measures.[50]

Although no country's soil is adequately secured, there are occasional signs of hope. One is the trend toward reduced tillage in the United States, triggered by farmers' desire to reduce fuel consumption and operating costs. So far the farmers turning to reduced tillage are not usually the ones with the most rapidly eroding soils. But reduced tillage may become an economically attractive first line of defense against erosion, particularly given the high cost of constructing terraces and adopting long-term rotations and other traditional approaches to soil conservation.

Another hopeful example is Kenya's ambitious national soil conservation program. Less than a decade old, Kenya's program shows not only that conservation is compatible with small-scale farming and a large rural population, but also that conservation improvements can boost farmers' incomes. A similar approach could work in many Third World countries.

Yet another encouraging development is the response to the erosion threat by the international scientific community, as evidenced by several recent conferences and specially commissioned studies. The International Congress of Soil Science, which met in New Delhi in 1982, focused on the need for a world soils policy. In early 1983, the Soil Conservation Society of America convened an international conference on soil erosion and conservation; some 145 scientists from around the world presented papers. And the American Society of Agronomy recently published proceedings of an international symposium on soil erosion in the tropics.[51]

In the absence of successful efforts to stem the loss of topsoil, the social effects of erosion will probably first be seen in Africa, in the form of acute food shortages and higher mortality rates, particularly for infants. Africa's record population growth and rampant soil erosion, and the absence of an effective response to either, combine to ensure that the continent will be at the forefront of this unfolding global crisis. What is at stake is not merely the degradation of soil, but the degradation of life itself.

Historically, soil erosion was a local problem. Individual civilizations whose food systems were undermined in earlier times declined in isolation. But in the integrated global economy of the late twentieth century, food—like oil—is a global commodity. The excessive loss of topsoil anywhere ultimately affects food prices everywhere.

Notes

1. Anson R. Bertrand, "Overdrawing the Nation's Research Accounts," *Journal of Soil and Water Conservation*, May/June 1980.

2. Population growth calculated from Population Reference Bureau, *1984 World Population Data Sheet* (Washington, D.C.: 1984).

3. Authors' estimates based on Table 10-9 "World Soil Resources and Excessive Soil Loss, 1980, with Projections to 2000," from Lester R. Brown et al., *State of the World 1984* (New York: W.W. Norton & Co., 1984).

4. U.S. Department of Agriculture (USDA), Soil Conservation Service, and Iowa State University Statistical Laboratory, *Basic Statistics 1977 National Resources Inventory*, Statistical Bulletin No. 686 (Washington, D.C.: 1982); USDA, Soil Conservation Service, "Preliminary 1982 National Resources Inventory," (Washington, D.C.: unpublished printout, April 1984).

5. P. O. Aina et al., quoted in R. Lal, "Effective Conservation Farming Systems for the Humid Tropics," in American Society of Agronomy, *Soil Erosion and Conservation in the Tropics*, ASA Special Publication No. 43 (Madison, Wisc.: 1982).

6. S. A. El-Swaify and E. W. Dangler, "Rainfall Erosion in the Tropics: A State-of-the-Art," in American Society of Agronomy, *Soil Erosion*.

7. These and other conservation practices introduced in the thirties are discussed in Donald Worster, *Dust Bowl* (New York: Oxford University Press, 1979), especially in Chapter 14, "Making Two Blades of Grass Grow."

8. Figures on summer fallow from U.S. Department of Agriculture (USDA), Economic Research Service, *Economic Indicators of the Farm Sector: Production and Efficiency Statistics, 1981* (Washington, D.C.: U.S. Government Printing Office, 1983).

9. Kenneth Grant, "Erosion in 1973-74: The Record and the Challenge," *Journal of Soil and Water Conservation*, January/February 1975; USDA, Soil Conservation Service, *Basic Statistics*.

10. USDA, Economic Research Service, "World Agriculture Outlook and Situation," Washington, D.C., June 1983.

11. Wouter Tims, *Nigeria: Options for Long-Term Development* (Baltimore, Md.: Johns Hopkins University Press, for the World Bank, 1974).

12. John M. Schiller et al., "Development of Areas of Shifting Cultivation in North Thailand 'Thai-Australia Land Development Project,'" in American Society of Agronomy, *Soil Erosion.*

13. R. F. Watters, *Shifting Cultivation in Latin America* (Rome: United Nations Food and Agriculture Organization, 1971).

14. Sheldon Judson, "Erosion of the Land, or What's Happening to Our Continents," *American Scientist*, July/August 1968.

15. Jiang Degi et al., "Soil Erosion and Conservation in the Wuding River Valley" (Beijing, China: Yellow River Conservancy Commission, April 1980), cited in El-Swaify and Dangler, "Rainfall Erosion in the Tropics"; U.S. Water Resources Council, *The Nation's Water Resources 1975-2000, Volume 2: Water Quantity, Quality and Related Land Considerations* (Washington, D.C.: U.S. Government Printing Office, 1978).

16. Ganges data from American Society of Agronomy, *Soil Erosion*; Mississippi data from U.S. Water Resources Council, *The Nation's Water Resources 1975-2000.*

17. J. M. Prospero et al., "Atmospheric Transport of Soil Dust from Africa to South America," *Nature*, February 12, 1981.

18. Josef R. Parrington et al., "Asian Dust: Seasonal Transport to the Hawaiian Islands," *Science*, April 8, 1983.

19. USDA, Soil Conservation Service, *Basic Statistics*, and "Preliminary 1982 National Resources Inventory."

20. *Ibid.*

21. K. G. Tejwani, Land Use Consultants International, New Delhi, private communication, July 3, 1983; Centre for Science and Environment, *The State Of India's Environment 1982* (New Delhi: 1982).

22. Quoted in Vera Rich, "Soil First," *Nature*, February 12, 1982.

23. Abandoned cropland figure cited is from U.S. Central Intelligence

Agency, *U.S.S.R. Agricultural Atlas* (Washington, D.C.: 1974); Thane Gustafson, "Transforming Soviet Agriculture: Brezhnev's Gamble on Land Improvement," *Public Policy*, Summer 1977.

24. P. Poletayev and S. Yashukova, "Environmental Protection and Agricultural Production," *Ekonomika Sel'skogo Khozyaystva* (Moscow), November 1978.

25. P. S. Tregubov, "Soil Rainstorm Erosion and Its Control in the U.S.S.R.," presented to the International Conference on Soil Erosion and Conservation, Honolulu, Hawaii, January 16-22, 1983; U.S. erosion rates are from USDA, "Preliminary 1982 National Resources Inventory."

26. National Soil Erosion-Soil Productivity Research Planning Committee, USDA, Science and Education Administration, "Soil Erosion Effects on Soil Productivity: A Research Perspective," *Journal of Soil and Water Conservation*, March/April 1981.

27. *Ibid.*

28. G. W. Langdale et al., "Corn Yield Reduction on Eroded Southern Piedmont Soils," *Journal of Soil and Water Conservation*, September/October 1979.

29. W. E. Larson, F. J. Pierce, and R. H. Dowdy, "The Threat of Soil Erosion to Long-Term Crop Production," *Science*, February 4, 1983.

30. Research by R. Lal of the International Institute of Tropical Agriculture, Ibadan, Nigeria, cited by El-Swaify and Dangler, "Rainfall Erosion in the Tropics." Even more pronounced yield reductions are investigated in J. S. C. Mbagwu, R. Lal, and T. W. Scott, "Effects of Desurfacing on Alfisols and Ultisols in Southern Nigeria I: Crop Performance," unpublished.

31. El-Swaify and Dangler, "Rainfall Erosion in the Tropics." The effects of environmental degradation on the lifespan of the Mangla Reservoir are discussed in Erik Eckholm, *Losing Ground: Environmental Stress and World Food Prospects* (New York: W.W. Norton & Co., 1976).

32. U.S. Agency for International Development, *Environmental and Natural Resource Management in Developing Countries: A Report to Congress* (Washington, D.C.: 1979).

33. Eckholm, *Losing Ground.*

34. B. B. Vohra, *Land and Water Management Problems in India,* Training Volume 8 (New Delhi: Ministry of Home Affairs, 1982).

35. Dr. Frank Wadsworth, "Deforestation—Death to the Panama Canal," in U.S. Department of State, Office of Environmental Affairs, *Proceedings of the U.S. Strategy Conference on Tropical Deforestation,* June 12-14, 1978 (Washington, D.C.: 1978).

36. A comprehensive introduction to the techniques and advantages of conservation tillage is provided in William A. Hayes, *Minimum Tillage Farming* and H. M. Young, *No-Tillage Farming* (bound in one volume) (Brookfield, Wisc.: No-Till Farmer, Inc., 1982).

37. Robert J. Gray, American Farmland Trust, statement before the Subcommittee on Conservation, Credit and Rural Development, Committee on Agriculture, U.S. House of Representatives, May 4, 1983.

38. Poletayev and Yashukova, "Environmental Protection."

39. National Soil Erosion-Soil Productivity Research Planning Committee, "Soil Erosion Effects on Soil Productivity."

40. The Erosion-Productivity Impact Calculator is described in USDA, Soil Conservation Service, National Bulletin No. 450-3-5, Washington, D.C., January 17, 1983. The Productivity Index (PI) is described in Larson, Pierce, and Dowdy, "The Threat of Soil Erosion to Long-Term Crop Production." International work on the PI is discussed in Frank R. Rijsberman and M. Gordon Wolman, eds., *Quantification of the effect of erosion on soil productivity in an international context* (Delft, The Netherlands: Delft Hydraulics Laboratory, 1984).

41. Roosevelt's personal commitment to soil conservation and particularly to the shelterbelt idea is recalled in Richard Strout, "Windbreaks: FDR Legacy," *Christian Science Monitor,* February 8, 1982.

42. "Soil Conservation in Kenya: An interview with Carl-Gosta Wenner," *Ambio,* Vol. 12, No. 6, 1983.

43. *Ibid.*

44. Swedish International Development Authority, *Soil Conservation in Kenya 1980 Review* (Stockholm: December 1980).

45. B. B. Vohra, "Managing India's Land and Water Resources," New Delhi, Ministry of Petroleum, August 1978.

46. Alfredo Sfeir-Younis, Agriculture and Rural Development Department, World Bank, private communication, May 26, 1983.

47. Authors' estimate of sugarcane acreage for alcohol production in Brazil based on James Bruce, "Brazil's Alcohol Export Prospects Shrink But Domestic Mart Continues Growth," *Journal of Commerce,* April 5, 1983. Authors' estimate of corn acreage for alcohol production in the U.S. based on data from Janet Livezey, Economic Research Service, USDA, private communication, August 15, 1983. R. Neil Sampson, "Energy: New Kinds of Competition for Land," presented to the 35th Annual Meeting of the Soil Conservation Society of America, Dearborn, Michigan, August 6, 1980.

48. Lloyd K. Fischer, discussion comments, in Harold G. Halcrow, Earl O. Heady, and Melvin L. Cotner, eds., *Soil Conservation Policies, Institutions and Incentives* (Ankeny, Iowa: Soil Conservation Society of America, 1982).

49. Ken Cook, "Surplus Madness," *Journal of Soil and Water Conservation,* January/February 1983.

50. American Farmland Trust, *Soil Conservation in America: What Do We Have to Lose?* (Washington, D.C.: 1984).

51. Symposia papers of the 12th International Congress of Soil Science, February 8-16, 1982, were published as *Desertification and Soils Policy* (New Delhi: Indian Society of Soil Science, 1982). Proceedings of the International Conference on Soil Erosion and Conservation, January 16-22, 1983, should be available in 1984; abstracts of papers presented at the conference can be obtained from the Department of Agronomy and Soil Science, University of Hawaii at Manoa, Honolulu, Hawaii. Proceedings of the ASA symposium were published in American Society of Agronomy, *Soil Erosion.*

LESTER R. BROWN is President of and a Senior Researcher with Worldwatch Institute. Formerly Administrator of the International Agricultural Development Service of the United States Department of Agriculture, he is the author of several books, including *World Without Borders*, *By Bread Alone*, *The Twenty-Ninth Day*, and *Building a Sustainable Society*, and co-author of *State of the World 1984*.

EDWARD C. WOLF, a Researcher with Worldwatch Institute, served as Project Assistant for *State of the World 1984*. He is a graduate of Williams College with a degree in Biology and Environmental Studies.

THE WORLDWATCH PAPER SERIES

No. of Copies

______ 1. **The Other Energy Crisis: Firewood** by Erik Eckholm.
______ 2. **The Politics and Responsibility of the North American Breadbasket** by Lester R. Brown.
______ 3. **Women in Politics: A Global Review** by Kathleen Newland.
______ 4. **Energy: The Case for Conservation** by Denis Hayes.
______ 5. **Twenty-two Dimensions of the Population Problem** by Lester R. Brown, Patricia L. McGrath, and Bruce Stokes.
______ 6. **Nuclear Power: The Fifth Horseman** by Denis Hayes.
______ 7. **The Unfinished Assignment: Equal Education for Women** by Patricia L. McGrath.
______ 8. **World Population Trends: Signs of Hope, Signs of Stress** by Lester R. Brown.
______ 9. **The Two Faces of Malnutrition** by Erik Eckholm and Frank Record.
______ 10. **Health: The Family Planning Factor** by Erik Eckholm and Kathleen Newland.
______ 11. **Energy: The Solar Prospect** by Denis Hayes.
______ 12. **Filling The Family Planning Gap** by Bruce Stokes.
______ 13. **Spreading Deserts—The Hand of Man** by Erik Eckholm and Lester R. Brown.
______ 14. **Redefining National Security** by Lester R. Brown.
______ 15. **Energy for Development: Third World Options** by Denis Hayes.
______ 16. **Women and Population Growth: Choice Beyond Childbearing** by Kathleen Newland.
______ 17. **Local Responses to Global Problems: A Key to Meeting Basic Human Needs** by Bruce Stokes.
______ 18. **Cutting Tobacco's Toll** by Erik Eckholm.
______ 19. **The Solar Energy Timetable** by Denis Hayes.
______ 20. **The Global Economic Prospect: New Sources of Economic Stress** by Lester R. Brown.
______ 21. **Soft Technologies, Hard Choices** by Colin Norman.
______ 22. **Disappearing Species: The Social Challenge** by Erik Eckholm.
______ 23. **Repairs, Reuse, Recycling—First Steps Toward a Sustainable Society** by Denis Hayes.
______ 24. **The Worldwide Loss of Cropland** by Lester R. Brown.
______ 25. **Worker Participation—Productivity and the Quality of Work Life** by Bruce Stokes.
______ 26. **Planting for the Future: Forestry for Human Needs** by Erik Eckholm.
______ 27. **Pollution: The Neglected Dimensions** by Denis Hayes.
______ 28. **Global Employment and Economic Justice: The Policy Challenge** by Kathleen Newland.
______ 29. **Resource Trends and Population Policy: A Time for Reassessment** by Lester R. Brown.
______ 30. **The Dispossessed of the Earth: Land Reform and Sustainable Development** by Erik Eckholm.
______ 31. **Knowledge and Power: The Global Research and Development Budget** by Colin Norman.
______ 32. **The Future of the Automobile in an Oil-Short World** by Lester R. Brown, Christopher Flavin, and Colin Norman.
______ 33. **International Migration: The Search for Work** by Kathleen Newland.
______ 34. **Inflation: The Rising Cost of Living on a Small Planet** by Robert Fuller.
______ 35. **Food or Fuel: New Competition for the World's Cropland** by Lester R. Brown.
______ 36. **The Future of Synthetic Materials: The Petroleum Connection** by Christopher Flavin.
______ 37. **Women, Men, and The Division of Labor** by Kathleen Newland.

_____ 38. **City Limits: Emerging Constraints on Urban Growth** by Kathleen Newland.
_____ 39. **Microelectronics at Work: Productivity and Jobs in the World Economy** by Colin Norman.
_____ 40. **Energy and Architecture: The Solar and Conservation Potential** by Christopher Flavin.
_____ 41. **Men and Family Planning** by Bruce Stokes.
_____ 42. **Wood: An Ancient Fuel with a New Future** by Nigel Smith.
_____ 43. **Refugees: The New International Politics of Displacement** by Kathleen Newland.
_____ 44. **Rivers of Energy: The Hydropower Potential** by Daniel Deudney.
_____ 45. **Wind Power: A Turning Point** by Christopher Flavin.
_____ 46. **Global Housing Prospects: The Resource Constraints** by Bruce Stokes.
_____ 47. **Infant Mortality and the Health of Societies** by Kathleen Newland.
_____ 48. **Six Steps to a Sustainable Society** by Lester R. Brown and Pamela Shaw.
_____ 49. **Productivity: The New Economic Context** by Kathleen Newland.
_____ 50. **Space: The High Frontier in Perspective** by Daniel Deudney.
_____ 51. **U.S. and Soviet Agriculture: The Shifting Balance of Power** by Lester R. Brown.
_____ 52. **Electricity from Sunlight: The Future of Photovoltaics** by Christopher Flavin.
_____ 53. **Population Policies for a New Economic Era** by Lester R. Brown.
_____ 54. **Promoting Population Stabilization: Incentives for Small Families** by Judith Jacobsen
_____ 55. **Whole Earth Security: A Geopolitics of Peace** by Daniel Deudney
_____ 56. **Materials Recycling: The Virtue of Necessity** by William U. Chandler
_____ 57. **Nuclear Power: The Market Test** by Christopher Flavin
_____ 58. **Air Pollution, Acid Rain, and the Future of Forests** by Sandra Postel
_____ 59. **Improving World Health: A Least Cost Strategy** by William U. Chandler
_____ 60. **Soil Erosion: Quiet Crisis in the World Economy** by Lester Brown and Edward Wolf.

_____ **Total Copies**

Single Copy—$4.00

Bulk Copies (any combination of titles)
2-5: $3.00 each 5-20: $2.00 each 21 or more: $1.00 each

Calendar Year Subscription (1984 subscription begins with Paper 58)
U.S. $25.00 _____

Make check payable to Worldwatch Institute
1776 Massachusetts Avenue NW, Washington, D.C. 20036 USA

Enclosed is my check for U.S. $ _____

name

address

city **state** **zip/country**